MW00490721

DECIMALS

GLOBE FEARON EDUCATIONAL PUBLISHER
A Division of Simon & Schuster
Upper Saddle River, New Jersey

Executive Editor: Barbara Levadi
Editors: Bernice Golden, Lynn Kloss, Bob McIlwaine, Kirsten Richert, Tom Repensek
Production Manager: Penny Gibson
Production Editor: Walt Niedner
Interior Design: The Wheetley Company
Electronic Page Production: The Wheetley Company
Cover Design: Pat Smythe

Reviewers:

Cheryl Miller, Mathematics Teacher
Forestville High School, Forestville, MD

Elliott Ringhel
Assistant Principal for Mathematics
Prospect Heights High School, Brooklyn, NY

Printed in the United States of America 1 2 3 4 5 6 7 8 9 10 99 98 97 96 95

ISBN 0-8359-1549-2

GLOBE FEARON EDUCATIONAL PUBLISHER
A Division of Simon & Schuster
Upper Saddle River, New Jersey

CONTENTS

TO THE STUDENT

Access to Math is a series of 15 books designed to help you learn new skills and practice these skills in mathematics. You'll learn the steps necessary to solve a range of mathematical problems.

LESSONS HAVE THE FOLLOWING FEATURES:

❖ Lessons are easy to use. Many begin with a sample problem from a real-life experience. After the sample problem is introduced, you are taught step-by-step how to find the answer. Examples show you how to use your skills.

❖ The *Guided Practice* section demonstrates how to solve a problem similar to the sample problem. Answers are given in the first part of the problem to help you find the final answer.

❖ The *Exercises* section gives you the opportunity to practice the skill presented in the lesson.

❖ The *Application* section applies the math skill in a practical or real-life situation. You will learn how to put your knowledge into action by using manipulatives and calculators, and by working problems through with a partner or a group.

Each book ends with *Cumulative Reviews*. These reviews will help you determine if you have learned the skills in the previous lessons. The *Selected Answers* section at the end of each book lists answers to the odd-numbered exercises. Use the answers to check your work.

Working carefully through the exercises in this book will help you understand and appreciate math in your daily life. You'll also gain more confidence in your math skills.

PLACE VALUE

decimal: a number with place values to the right of the ones place to express parts of one, such as tenths and hundredths

decimal point: a point used to separate the ones place from the tenths place

digit: any of the ten symbols, 0–9, that name a number

place value: a digit's value based on its position in a number

As she began her work with the decimal system, Lin made a useful discovery. She found that she uses **decimals** every day when she handles money. In the number below, a **decimal point** separates a whole number of dollars from parts of a dollar.

$ 110. 45
whole dollars   parts of a dollar

Lin knows that a penny is actually *one hundredth* of a dollar, so the 45 cents above can be called 45 *hundredths* of a dollar.

A **place-value** chart shows the value of each **digit** in a number, depending on its position in the number.

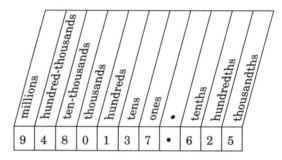

For example, the place-value chart above shows that the 1 is in the *hundreds* place. Therefore, its value is 1×100, or 100. The 8 is in the *ten-thousands* place. Its value is $8 \times 10,000$, or 80,000.

The digits to the right of the **decimal point** express parts of one. The 6 is in the *tenths place*, so its value is 6×0.1, or 0.6. The 2 is in the *hundredths place*, so its value is 2×0.01, or 0.02.

Guided Practice

1. Read the number in the place-value chart above.

   a. What digit is in the *millions* place? _____9_____

   b. What is the value of that digit?
     9 × 1,000,000, or 9,000,000

   c. What digit is in the *tens* place? _____

d. What is the value of that digit? _____

e. What digit is in the *thousandths* place?

_____

f. What is the value of that digit? _____

Exercises

Write the value of the 4 in each number.

2. 54, 912

3. 391.40

4. 567,014.11

_____ _____ _____

Write the value of the 9 in each number.

5. 8,342.09

6. 7,908,432.7

7. 4,598.22

_____ _____ _____

Write the value of the 5 in each number.

8. 37.005

9. 9,213.5

10. 627,398.459

_____ _____ _____

Application

COOPERATIVE
LEARNING

11. In the place value chart below, write the greatest number you can, using each digit from 0 to 9 only once. Compare your answer with a partner's and see who got the greatest number. How did you decide what number to write?

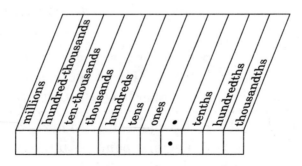

_____

_____

READING AND WRITING DECIMALS

Josh answered a help-wanted ad and was told that the delivery job paid "seven dollars and ninety cents per hour."

Numbers can be written in words as well as with digits. Numbers written with digits are in standard form. Because it's easier to write numbers in standard form, Josh writes down the salary as a decimal, "$7.90." Did he write the number correctly?

Look at the decimal on a place value chart.

Reminder

The "and" in a written or spoken number stands for the decimal point.

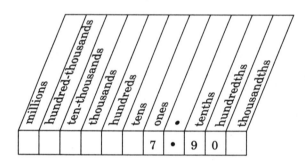

Josh's standard form of the salary is correct.

Guided Practice

1. Write 10.05 in words.

   a. Start with the places to the left of the decimal point. How would you say this part of the number? _____ten_____

   b. Look at the places to the right of the decimal point. How would you say this part of the number? _five hundredths_

   c. Now join the two parts of the number at the decimal point with the word *and*. Write the decimal.

   _____

Write each decimal in words.

2. 55.001

_____

3. 1.05

_____

4. 0.003

_____

Write each number in standard form.

5. Three hundred twenty-four thousand

_____

6. Two hundred sixty-two and forty-five thousandths

_____

7. Nine and seven thousandths

_____

Application

8. Read these numbers aloud to a friend and have him or her write the number in standard form. Then switch roles.

a. "fourteen dollars and seven cents"

Standard form: _____

b. "three hundred fifty and one tenth"

Standard form: _____

c. "fifteen and five hundredths"

Standard form: _____

d. "one thousand and one hundred twenty-five thousandths"

Standard form: _____

DECIMALS ON THE NUMBER LINE

Students in Leona's math class are comparing how many hours they work each week at their after-school jobs. Leona writes their hours on a number line. By using a number line, she quickly can compare the numbers. Ann Lih works 5 hours, Kyle works 10 hours, and Tony works 7 hours. Here's what Leona recorded:

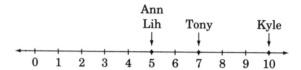

If Laquita works 8.5 hours, how can Leona mark this number on the number line? The number line below shows each whole number divided into ten equal parts, or **tenths**.

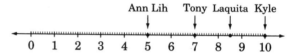

Laquita's eight and five tenths hours are recorded correctly here.

Guided Practice

1. The number line below is divided into ten equal parts. Each part is one tenth of the space between 0 and 1. The first marks correspond to 0.1, 0.2, and 0.3. Complete the numbering.

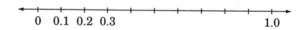

a. On the number line locate the point that corresponds to the decimal 0.45. The decimal 0.45 is halfway between 0.4 and 0.5. Label the point *A*.

b. Locate and label 0.35 on the number line.

c. Locate and label 0.75 on the number line.

Locate a point for each number on the number line below. Label points that are between the marks on the line.

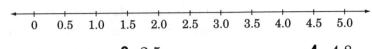

```
+----+----+----+----+----+----+----+----+----+----+---->
0   0.5  1.0  1.5  2.0  2.5  3.0  3.5  4.0  4.5  5.0
```

2. 2   **3.** 2.5   **4.** 4.8

5. 1.3   **6.** 3.7   **7.** 0.75

The chart below shows part of a scale from a Fahrenheit thermometer. It is used to record a person's temperature.

102
101
100
99
98

8. Write in standard form the temperature "one hundred and two tenths." This is also read "one hundred point two." _____

9. The normal temperature for a healthy person is 98.6°F. Draw an arrow to show where this is on the thermometer.

10. The thermometer shows the temperature of a person with a fever. What is the temperature? _____

Enter these numbers on a calculator. Record what appears on the display.

11. three thousand five and twenty hundredths _____

12. fifteen and six tenths _____

13. twenty-five and one hundred fifty-five thousandths _____

COMPARING AND ORDERING DECIMALS

Hector, a stockboy, weighs packages for shipping. He stacks them so that the heaviest package is on the bottom and the lightest package is on the top. He compares the weights of three packages.

Package A	Package B	Package C
3.068 lb	3.6 lb	3.079 lb

To find out which package weighs the most, compare the decimals.

List the numbers so that the decimal points are lined up. Label the place value columns. Then compare the digits in each column, from left to right.

Reminder

The symbol < means *is less than* and the symbol > means *is greater than*.

Ones	Tenths	Hundredths	Thousandths
3 .	0	6	8
3 .	6		
3 .	0	7	9
The ones value for each number is the same.	0.6 > 0.0 so 3.6 is the greatest number.	0.079 > 0.068 so 3.068 is the least number.	

The numbers in order from least to greatest are 3.068, 3.079, 3.6.

Package B weighs the most and should be at the bottom of the stack.

Guided Practice

1. Compare the decimals 450.17, 450.1, and 450.09. Use < or >.

   a. Compare hundreds. Are the digits the same?

   _____ yes _____

   b. Compare tens. Are the digits the same?

   _____

   c. Compare ones. Are the digits the same?

   _____

d. Compare tenths. 1 _____ 0

e. Compare hundredths. 17 _____ 10

f. Order the decimals from least to greatest.

_____

Exercises

Order decimals from least to greatest.

2. 58.83   58.71   55.84

_____

3. 39.09   39.6   32.98

_____

4. 26.08   29.14   29.31

_____

5. 15.75   5.75   57.5

_____

6. 1.14   1.22   0.76

_____

7. 429.6   429.34   429.01

_____

Compare. Write <, >, or =.

8. 89.099 _____ 89.2

9. 201.5 _____ 210.66

10. 400.11 _____ 400.3

11. 56,781.22 _____ 56,781.22

Application

12. Describe how you would compare the decimals 5.5 and 5.25.

_____

_____

_____

ROUNDING DECIMALS

Vocabulary

round: drop or replace digits with zeros so that the number is easier to use

Ibrahim is trying to talk David into running a 5-kilometer footrace through Chicago's Lincoln Park. Although David jogs, he has never run a 5km race before. He wonders about how many miles 5 kilometers is.

One kilometer is equal to 0.6214 miles. For an exact answer, David could use a calculator to multiply the number of kilometers, 5, by 0.6214. Or he could **round** this decimal and use mental math to estimate the answer.

Round 0.6214 to the nearest tenth.

To round a decimal, look at the digit in the place value to be rounded. In this case, it is six tenths.

$$0.\underline{6}214$$

Next, look at the digit to the right of the digit to be rounded.

- If this digit is less than 5, the digit to be rounded remains the same.

- If this digit is 5 or more, add 1 to the digit to be rounded.

$$0.\underline{6}214$$

Because 2 is less than 5, the 6 doesn't change.

When rounding to the right of the decimal point, all the digits to the right of the rounded digit can be dropped. (When rounding money amounts to the nearest 10 cents, drop the rounded digit and add zero. So $2.48 becomes $2.50.)

So, 0.6214 rounded to the nearest tenth is 0.6.

David can multiply 5 by 0.6 mentally to find out about how many miles 5 kilometers is.

$$5 \times 0.6 = 3.0 = 3$$

David estimates that a 5-kilometer race is about 3 miles.

1. Round 155.749 to the nearest hundredth.

   a. Which digit is to be rounded? _____4_____

   b. Which digit is to the right of the digit to be rounded? _____9_____

   c. Is this digit less than, equal to, or greater than 5? _____

   d. Write 155.749 rounded to the nearest hundredth. _____

Exercises

Round each number to the underlined place value.

2. 3.58_7_2

_____

3. $67._0_5

_____

4. 45.6_7_4

_____

5. 99.019_1_8

_____

6. 376,_2_11

_____

7. $75_2_.90

_____

8. The speed for the Indianapolis 500 winner in 1990 was 185.987 miles per hour, a record set by Arie Luyendyk.

   a. What is this speed rounded to the nearest mile per hour?

   _____

   b. What is the speed rounded to the nearest tenth of a mile per hour?

   _____

Application

9. Empty the change from your pocket, your change purse, or a money jar.

   a. How much money is there rounded to the nearest dime?

   _____

   b. How much money is there rounded to the nearest dollar?

   _____

   c. Which rounded amount is more accurate?

   _____

FINDING EQUIVALENT DECIMALS AND FRACTIONS

At the end of every week, Carlos never can figure out where all his money went. A friend suggested that he record his spending, then make out a weekly budget. The table below shows how much of every dollar Carlos spends on different "necessities."

Entertainment and fast food	$0.50
Clothes	$0.25
Gas	$0.15
School supplies	$0.10

When discussing how money is spent, you can often use fractions.

Example 1

What fraction of Carlos's money is spent on entertainment and fast food?

To answer this question, write a fraction that is **equivalent** to 0.50.

Recall that 0.50 means 50 hundredths.

Write 50 hundredths as a fraction.

$$0.50 = \frac{50}{100}$$

Simplify $\frac{50}{100}$.

$$\frac{50}{100} \xrightarrow{\div 50} = \frac{1}{2} \xleftarrow{\div 50}$$

Reminder

You can simplify a fraction if you can divide the numerator and denominator by the same number.

Carlos spends one half of his money on fast food and entertainment.

Example 2

What fraction of Carlos's money is spent on school supplies?

Write 0.10 as a fraction.

$$0.10 = \frac{10}{100} = \frac{1}{10}$$

Carlos spends one tenth of his money on school supplies.

1. Change 0.25 to a fraction.

    a. Write 25 hundredths as a fraction. ____$\frac{25}{100}$____

    b. Simplify the fraction. ____$\frac{25}{100} = \frac{1}{4}$____

2. Change 0.625 to a fraction.

    a. Write 625 thousandths as a fraction. _____

    b. Simplify the fraction. _____

Exercises

Write each decimal as a fraction in simplest form.

3. 0.10   4. 0.15   5. 0.01

_____   _____   _____

6. 0.025   7. 0.375   8. 0.001

_____   _____   _____

9. 0.05   10. 0.75   11. 0.123

_____   _____   _____

Change the following decimals to fractions with the same denominator. Then simplify the fractions.

12. 0.05   13. 0.5   14. 0.005

_____   _____   _____

_____   _____   _____

Application

15. Do you know your spending habits? Fill in the chart below. You can use the same categories as Carlos used or different ones.

	$
	$
	$
	$

FINDING EQUIVALENT FRACTIONS AND DECIMALS

Jessica ordered $\frac{1}{4}$ pound of chef salad at the deli counter. When the salesperson weighed the salad, the scale displayed 0.25 pound. Did Jessica get the right amount of salad? How do you know?

Sometimes fractions are used to represent numbers less than 1, and sometimes decimals are used. You can change any fraction to an equivalent decimal by dividing the numerator by the denominator.

$$\textit{fraction} \longrightarrow \textit{decimal}$$

$$\frac{1}{4} = \begin{array}{r} 0.25 \\ 4\overline{)1.00} \end{array}$$

So, 0.25 pound is the same as $\frac{1}{4}$ pound.

Sometimes when you change a fraction to an equivalent decimal, the division does not come out evenly. For example, change $\frac{1}{3}$ to a decimal.

$$\textit{fraction} \longrightarrow \textit{decimal}$$

$$\frac{1}{3} = \begin{array}{r} 0.33\frac{1}{3} \\ 3\overline{)1.00} \\ \underline{9} \\ 10 \\ \underline{9} \\ 1 \end{array}$$

Guided Practice

1. To find an equivalent decimal for a fraction, divide the numerator by the ___denominator___ .

2. Write $\frac{4}{5}$ as a decimal. _____

3. Write $\frac{5}{6}$ as a decimal. _____

Write each fraction as a decimal.

4. $\frac{3}{4}$ **5.** $\frac{5}{8}$ **6.** $\frac{1}{2}$

    _____   _____   _____

7. $\frac{1}{6}$ **8.** $\frac{3}{10}$ **9.** $\frac{2}{3}$

    _____   _____   _____

Write an equivalent fraction or a mixed number for each decimal.

10. $0.543 =$ _____   **11.** $0.8 =$ _____   **12.** $2.06 =$ _____

Application

13. Create your own "concentration" game to practice your decimal-fraction equivalence skills. Follow these steps.

    a. On index cards, write the following numbers—one number on each card.

$\frac{1}{2}$	$\frac{1}{3}$	$\frac{2}{3}$	$\frac{1}{4}$	$\frac{3}{4}$	$\frac{1}{5}$	$\frac{2}{5}$	$\frac{3}{5}$
0.50	0.33	0.67	0.25	0.75	0.2	0.4	0.6

    b. Mix up the cards and lay them face down on a table.

    c. Take turns with a partner. Turn over one card. See if you can turn over the card that shows its correct equivalent.

ESTIMATING DECIMAL SUMS AND DIFFERENCES

Vocabulary

estimate: to find a value that is close to an exact number

difference: in subtraction, the amount by which one number is less than another

sum: the result when two or more numbers are added together

Reminder

To round a decimal, look at the digit to the right of the digit to be rounded. If it is less than 5, round down. If it is 5 or greater, round up.

Example 1

Anthony has $33.25 in his pocket. He wants to buy a CD that costs $13.79, a box of blank tapes that costs $8.90, and a poster that costs $4.50. Does he have enough money?

You can **estimate** the **sum** by rounding all decimals to the nearest whole number and adding.

$$\$13.79 \rightarrow rounds\ to \rightarrow \$14$$
$$8.90 \rightarrow rounds\ to \rightarrow 9$$
$$4.50 \rightarrow rounds\ to \rightarrow \underline{\quad 5}$$
$$\$28$$

Anthony's purchases will be about $28. He does have enough money.

Example 2

While standing in the checkout line, Anthony remembers that he owes his sister $6.79 for a book. If he spends about $28 of the $33.25 in his pocket, can he also pay back his sister?

You can estimate the **difference** by rounding the decimals to the nearest whole number and subtracting.

$$33.25 \rightarrow rounds\ to \rightarrow \begin{array}{r} \$33 \\ -28 \\ \hline \$\ 5 \end{array}$$

He will have about $5 left, not enough to pay back his sister.

Guided Practice

1. Estimate whether $50 will be enough to pay for three shirts that cost $19.85, $15.29, and $21.99.

   a. $19.85 rounds to _____$20_____.

   b. $15.29 rounds to _____.

c. $21.99 rounds to _____.

d. Estimate the sum. _____.

e. Is $50 enough money to pay for the shirts?

_____

2. Benjamin buys school supplies for $12.88. Use mental math to estimate Benjamin's change from a $20 bill.

a. $12.88 rounds to _____.

b. Subtract to estimate Benjamin's change.

_____

Exercises

Round decimals to the nearest whole number. Estimate the sum or difference.

3. 16.3 + 40.7

_____

4. 92.5 − 71.3

_____

5. 125.58 − 24.62

_____

6. $18.92
   25.17
   + 8.75
   _____

7. $99.25
   −62.78
   _____

8. 168.12
   29.93
   +45.77
   _____

Application

Use mental math to estimate whether the purchases listed can be paid for with $25. Write the estimate. Then write *yes* or *no*.

9. 2 pairs of socks at $2.99 each, and a belt for $19.59

_____

10. 3 cassettes at $7.95, $8.99, and $6.55

_____

ADDING DECIMALS

Curtis delivers office supplies for the Rodex Manufacturing Company. A recent order included stationery for $98.42, envelopes for $57.85, labels for $24.95, and file folders for $36.75. He needs to find a total for the order so that he can get a check from the customer.

To find the total, write the numbers with the decimal points lined up. Add as you would with whole numbers.

$$\begin{array}{r} \$98.42 \\ 57.85 \\ 24.95 \\ \underline{36.75} \\ \$217.97 \\ \uparrow \end{array}$$

Bring the decimal point directly down in the answer.

The total for the order is $217.97.

Guided Practice

Reminder

You can write zeros after the decimal points as placeholders so that all of the numbers have the same number of decimal places.

1. A company that makes parts to hold computer chips must cut pieces of metal to exact lengths. Find the total length of three pieces of metal that measure 3.8 cm, 9.356 cm, and 4.07 cm.

   a. Line up the decimal points. Write zeros after the decimal point, as needed.

   $$3.800$$

   $$9.356$$

   $$+ \underline{}$$

   b. Add the three decimals. _____

Estimate each sum. Then line up the decimal points and add.

2. Add: $42.86, $206.90, $3.04

Estimate: _____

Answer: _____

3. Add: $560, $1,209.47, $35.56

Estimate: _____

Answer: _____

4. Add: 43.8, 760.02, 6.798

Estimate: _____

Answer: _____

5. Add: 5.006, 442.17, 587.408

Estimate: _____

Answer: _____

6. Add twenty-three and fifty-six hundredths plus five and one hundred twenty-nine thousandths. _____

Application

 Use the prices given and a calculator to find each total cost.

Apples: $0.89 per pound

Pears: $1.49 per pound

Peppers: $2.59 per pound

Green beans: $1.79 per pound

7. 2 pounds of apples and 1 pound of pears

_____

8. 2 pounds of peppers and 1 pound of green beans

_____

9. 2 pounds of pears and 2 pounds of peppers

_____

10. 3 pounds of green beans and 2 pounds of apples

_____

ADDING WITH METRIC MEASURES

Vocabulary

metric measure: a system of measuring based on the decimal system

meter: the standard metric unit of length

kilometer: 1,000 meters

centimeter: 0.01 meters

A band is setting up its equipment on a big outdoor stage. If the stage is 3 meters high and the speaker cabinets are 85 centimeters tall, how high off the ground are the tops of the speakers?

To find the total height, you have to add meters and centimeters together. **Meters, kilometers,** and **centimeters** are units of measure in the **metric system.** To add meters and centimeters, change meters to centimeters.

There are 100 centimeters in 1 meter. Change 3 meters to an equivalent number of centimeters.

$$3 \ m \ = \ 3 \ \times \ 100 \ = \ 300 \ cm$$

Then add.

$$300 \ cm \ + \ 85 \ cm \ = \ 385 \ cm$$

The tops of the speakers are 385 cm, or 3.85 m, above the ground.

When changing from a *larger* unit, such as meters, to a *smaller* unit, such as centimeters, there will be more smaller units, so you multiply.

Guided Practice

Gary drove 4.25 kilometers before his car broke down. He had to walk 135 meters to reach a service station that had a tow truck.

1. Find the total distance Gary traveled.

   a. Change kilometers to meters.

   1 km = 1,000 m, so

   4.25 km = 4.25 × __1,000__ = _____ m

   b. Add _____ m + __135__ m = _____ m.

   c. The total distance Gary traveled was _____.

Add the following. Be careful with the units of measure.

2. 12.43 m, 9.25 m, 75 cm

_____

3. 1 m and 46 cm

_____

4. 2.5 m and 50 cm

_____

5. 5 cm, 8.5 cm, 2 m

_____

6. Add 3.7 km and 0.2 km. Express your answer in kilometers, then in meters.

a. _____ km

b. _____ m

7. On a county road map, 1 centimeter represents a distance of 2 kilometers on the road. What distance does 3.8 centimeters on the map represent?

_____

8. Sarah's family is planning a trip with stops in several towns. They use the same map as in Exercise 7. The distances between towns on the map are 9.3 cm, 5.8 cm, 8.1 cm, and 7.6 cm.

a. What is the total distance on the map? _____

b. What is the total distance on the road? _____

9. A garage was built on an oddly shaped four-sided lot. The sides measured 17 m, 900 cm, 18.5 m, and 700 cm. Find the lot's perimeter. _____

Application

10. The metric system is used in all the countries of the world except the United States. Should we adopt the metric system here? Why or why not?

✎ _____

_____

_____

SUBTRACTING DECIMALS

Vocabulary

balance: the amount of money that remains in an account after all transactions and charges have been deducted

withdraw: to take money out of the bank

Kathy's bank statement shows an account **balance** of $295.75. She **withdraws**, or takes out, $123.75. What is her new account balance?

To subtract decimals, line up the decimal points and subtract as you would with whole numbers. Remember to bring the decimal point directly down in your answer.

$295.75 Balance

−123.75 Withdrawal

$172.00 New balance

Kathy's balance is $172.

The next week, Kathy withdraws $95.27. What is the new balance?

Write $172 as $172. and line up the decimal points. Write two zeros as placeholders in the top number. Regroup and subtract.

$172.00

−95.27

$ 76.73

Kathy's new balance is $76.73.

Guided Practice

1. Alan has a bank balance of $405.08. He withdraws $99.85. Find his new balance.

   a. Write the balance. $405.08

   b. Write the amount withdrawn.
   Line up the decimal points.   − _____

   c. Subtract to find the new balance. $ _____

Find each difference.

2. $497.14
 −324.50

3. 210
 −172.2

4. 500.999
 −402.35

Estimate the difference. Then line up the decimal points and subtract.

5. Take $560.45 from $908.60.

Estimate: _____

Answer: _____

6. From $100, subtract $6.35.

Estimate: _____

Answer: _____

7. Find the difference between 88.71 and 20.

Estimate: _____

Answer: _____

8. Subtract 90.7 from 150.

Estimate: _____

Answer: _____

Application

COOPERATIVE LEARNING

Work with a partner.

9. In your local newspaper, find a page or two of advertisements for clothing, stereo equipment, or any other merchandise that appeals to you. Imagine that you have $1,000 to spend. Make a shopping list for your partner. Use your estimation skills to come as close to $1,000 as you can, without going over that amount. Then, your partner can practice his or her decimal skills by subtracting the cost of each of the chosen items from $1,000. Good luck and happy shopping!

_____

_____

_____

SUBTRACTING WITH METRIC MEASURES

To make posters advertising the student council meeting, Juanita bought poster board that measures 1 meter by 2 meters. It turns out that the maximum size the school will allow is 100 centimeters by 150 centimeters. How much will Juanita have to trim from the length of her poster board?

To solve a problem like this, you need to be able to change units of measure to other units—in this case, meters to centimeters.

Change 2 meters to an equivalent number of centimeters.

$$2 \ m = 2 \times 100 = 200 \ cm$$

Then subtract.

$$\begin{array}{r} 200 \ cm \\ -150 \ cm \\ \hline 50 \ cm \end{array}$$

Juanita will have to trim 50 cm, or 0.5 m, from the length of the poster board.

When changing from a *larger* unit, such as meters, to a *smaller* unit, such as centimeters, there will be more smaller units, so you multiply.

$$5 \ km = ? \ m$$
$$5 \ km = 5 \times 1,000 \ m$$
$$= 5,000 \ m$$

When changing from a *smaller* unit to a *larger* unit, there will be fewer units, so you divide.

$$500 \ cm = ? \ m$$
$$500 \ cm \div 100 = 5 \ m$$

Reminder
1 m = 100 cm
1 km = 1,000 m

1. Sarah lives 3.4 kilometers from work. One day she walked 800 meters and then got a ride. How far did the ride take her?

   a. Change kilometers to meters.

   1 km = 1,000 m, so 3.4 km = _____3,400_____ m

   b. Subtract: _____ m

   − _____ m

   _____ m

   c. The ride took her _____ m, or _____ km.

Exercises

Find each difference.

2. From 25.4 km, subtract 8.7 km.

   _____

3. Subtract 304.6 m from 500 m.

   _____

4. From 4 m, subtract 56 cm.

   _____

5. Subtract 1,230 m from 2.5 km.

   _____

6. Subtract 4.5 km from 4.6 km. Express your answer in meters, then in kilometers.

   a. _____ m

   b. _____ km

Application

7. For this activity, you'll need about 100 standard-size paper clips. Use these paper clips to make a floor plan of your apartment or house. Here are some tips.

   • Each paper clip is equivalent to 1 meter in length.

   • Estimate the sizes of your rooms. Remember that 1 meter is a little bit longer than 1 yard, and a little bit longer than 3 feet.

8. Describe your floor plan, using metric measures.

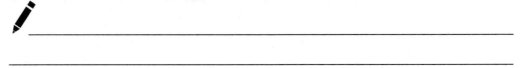

_____

_____

ESTIMATING DECIMAL PRODUCTS

Vocabulary

product: the result when two numbers are multiplied

area: the amount of space covered by a flat surface, such as a floor or a wall; area is measured in square units

Example 1

The ninth grade at Wood School charged $4.75 for a basic car wash at its annual spring fundraiser. The class washed 110 cars that day. Approximately how much money did the class raise?

The word *approximately* is a signal to estimate an answer. Round the numbers in the problem to get numbers that are easier to work with.

To find an answer to the problem above, you need to find the **product** of the cost of each car wash and the number of cars.

$4.75 rounds to $5   110 rounds to 100

100 × $5 = $500

The class raised approximately $500.

Example 2

The Wood School ninth grade plans to use its earnings to put a new surface on the indoor gym area. The floor measures 19.9 meters by 10.1 meters. Estimate the **area** of the floor.

To estimate the area, round the decimals to the nearest whole number. Then multiply length by width.

19.9 rounds to 20

10.1 rounds to 10

20 m (length) × 10 m (width) = 200 square meters or 200 m2

The area of the floor is about 200 square meters.

Guided Practice

1. The cost to pave a playground is $3.20 per square yard. The playground is 24.8 yards long and 15.2 yards wide.

a. Round all decimals to the nearest whole number.

$3.20 rounds to _$3_   24.8 rounds to _25_
15.2 rounds to ____

b. Estimate the area of the playground in square yards. _____

c. Estimate the cost of paving the playground.
_____

**Estimate the area of rectangles with the following dimensions.
Write in the appropriate square units.**

2. 14.9 ft by 27.3 ft

_____

3. 7.5 m by 8.4 m

_____

4. 7.1 in. by 6.3 in.

_____

5. 400.5 mi by 537 mi

_____

6. 4.5 cm by 3.5 cm

_____

7. 6.7 m by 6.8 m

_____

Application

8. To get a good idea of how area is really measured, try this activity with a pair of scissors, a ruler, paper, and a pencil.

a. With a ruler, measure a square 1 inch by 1 inch, and cut it out with scissors.

b. Using whatever method seems best to you, cut out 100 of these square inches.

c. Now use these square inches to cover the surface area of at least three square or rectangular objects, such as this book, a notebook, or a desktop. Count squares to find the area. Then find the area by multiplying the number of squares in the length by the number of squares in the width. List the objects and their areas below.

Object _____   Area _____

Object _____   Area _____

Object _____   Area _____

Object _____   Area _____

Object _____   Area _____

MULTIPLYING BY POWERS OF 10

Vocabulary

power: the product of a given number of identical factors, that is, whole numbers

liter: basic unit of capacity in the metric system

centiliter: 0.01 liter

deciliter: 0.10 liter

A **power** of 10, such as 10^5 (ten to the fifth power), means the factor 10 is repeated five times and there are five zeros in the product.

$$10^1 = 10$$
$$10^2 = 10 \times 10 = 100$$
$$10^3 = 10 \times 10 \times 10 = 1,000$$
$$10^4 = 10 \times 10 \times 10 \times 10 = 10,000$$
$$10^5 = 10 \times 10 \times 10 \times 10 \times 10 = 100,000$$

When you multiply a number such as 7.14 by 10 and by 100, you get the following results.

$$
\begin{array}{r}
7.14 \\
\times 10 \\
\hline
000 \\
7140 \\
\hline
71.40
\end{array}
\qquad
\begin{array}{r}
7.14 \\
\times 100 \\
\hline
000 \\
0000 \\
\hline
71400 \\
714.00
\end{array}
$$

Only the position of the decimal point differs in the two answers.

To multiply a number by 10, move the decimal point one place to the right. Notice that 10 has one zero. To multiply a number by 100, move the decimal point two places to the right. Notice that 100 has two zeros. You can use this pattern to predict that multiplying 7.14 by 1,000 will result in 7,140.

You now have a rule for multiplying by a power of 10: move the decimal point to the right as many places as there are zeros in the power of 10.

$$12.758 \times 100 = 12.7\,5\,8 = 1,275.8$$

Write zeros to the right of the number if you do not have enough places to move the decimal point.

$$0.943 \times 10,000 = 0.9\,4\,3\,0 = 9,430$$

Complete these exercises by placing the decimal point in the product.

1. 3.71 × 10 = ___37.1___

2. 0.024 × 1,000 = _____

3. 12.8 × 1,000 = _____

4. 619.5 × 100 = _____

Exercises

Multiply.

5. 0.03 × 1,000 _____

6. 0.00246 × 10,000 _____

7. 67.8 × 10 _____

8. 5.4 × 100 _____

9. 0.036 × 100 _____

10. 812.94 × 10 _____

11. Marge is buying 1,000 balloons for the school fair. At $0.29 each, how much will the balloons cost?

_____

12. The glass container on a blender holds 1.44 **liters** of a liquid. How many **centiliters** is this? How many **deciliters**?

_____

Application

13. Mathematicians and scientists use powers of 10 to express very, very large numbers. For example, some demographers estimate that the world population in the year 2000 will be 7×10^9. Astronomers calculate the distance the planet Neptune is from the sun as 4.5×10^9 miles. This kind of number expression is called scientific notation.

Why might people use scientific notation instead of writing out actual numbers? Write a short paragraph giving your ideas. As an example, try writing out one of the numbers above.

_____

_____

_____

_____

MULTIPLYING DECIMALS BY WHOLE NUMBERS

spreadsheet: a table with rows and columns of spaces called cells, containing either a number or a formula in each cell

Reminder

A formula is an equation that states a mathematical rule.

Carmina earns $4.85 per hour working at The Hamburger Shop on weekends. She works three hours on both Saturday and Sunday. To find her weekly pay, Carmina multiplies $4.85 times 6.

When multiplying a decimal by a whole number, multiply as you would with whole numbers. Then count the number of decimal places in the factors to determine where to place the decimal point in the product.

$$\begin{array}{r} \$4.85 \\ \times \quad 6 \\ \hline \$29.10 \end{array}$$

$4.85   (2 places after decimal point)

× 6

$29.10   (2 places after decimal point)

The Hamburger Shop uses a computer **spreadsheet** to save time calculating its weekly payroll. Each box in the spreadsheet is called a cell. Each cell is named by a column letter and a row number, such as A1 or D3. In each cell, the payroll clerk enters either a number or a formula.

	A	B	C	D
1				
2				
3				

The payroll clerk for The Hamburger Shop enters pay rates for all employees in Column A, hours worked each day in Column B, and days worked per week in Column C. In cell D1, she enters A1*B1*C1. The * means multiplication, so the computer multiplies the numbers in cells A1, B1, and C1, and puts the product in D1.

Suppose the payroll clerk enters Carmina's data in row 7. What number will the computer print in cell D7?

	A	B	C	D
7	4.85	3	2	A7 * B7 * C7

In cell D7, the computer would multiply the number in cell A7 ($4.85) times the number in cell B7 (3) times the number in cell C7 (2). So it would print $29.10 in cell D7, Carmina's weekly pay.

1. Place the decimal point in each answer. Write zeros as placeholders where necessary.

    a. 0.0004
    $\times$ 7
    —> 0.0028

    b. 0.031
    $\times$ 6
    186

    c. 4.65
    $\times$ 8
    3720

Exercises

Multiply.

2. 0.0032
$\times$ 6

3. 0.0571
$\times$ 3

4. 0.784
$\times$ 9

5. 4.85
$\times$ 6

6. 23.6
$\times$ 7

7. 14.032
$\times$ 8

8. 0.53
$\times$ 7

9. 5.85
$\times$ 8

10. 72.3
$\times$ 9

11. Suppose the number in cell A8 is 5.25 and the number in B8 is 4. If the payroll clerk enters A8 * B8 in cell C8, what number will appear in cell C8? _____

Application

12. Use your calculator to find the numbers in cells D1 through D5 on the spreadsheet.

	A	B	C	D
1	6.25	7	5	A1*B1*C2
2	4.75	8	3	A2*B2*C2
3	5.10	5	5	A3*B3*C3
4	3.90	4	4	A4*B4*C4
5	7.15	8	5	A5*B5*C5

a. cell D1 = _____

b. cell D2 = _____

c. cell D3 = _____

d. cell D4 = _____

e. cell D5 = _____

MULTIPLYING TWO DECIMALS

In sewing class, Anthony cut a piece of fabric that measured 1.5 yards by 2.5 yards. How many square yards of fabric did he have?

Reminder

Area is measured in square units.

To find the area, you multiply the length by the width.

$$1.5 \text{ yards} \times 2.5 \text{ yards} = ?$$

When you multiply a decimal by a decimal, just multiply as you would with whole numbers. Then add up the number of places after the decimal points in the factors to determine where the decimal point should go in the product.

$$
\begin{array}{rl}
1.5 & \text{(1 place after decimal point)} \\
\underline{\times\ 2.5} & \text{(1 place after decimal point)} \\
75 & \\
\underline{300} & \\
3.75 & \text{(2 places after decimal point)}
\end{array}
$$

Anthony had 3.75 square yards of fabric.

Sometimes you may have to write zeros as placeholders in order to place the decimal point correctly.

$$
\begin{array}{rl}
1.043 & \text{(3 places after decimal point)} \\
\underline{\times\ 0.0073} & \text{(4 places after decimal point)} \\
3129 & \\
\underline{73010} & \\
0.0076139 & \text{(7 places after decimal point)}
\end{array}
$$

Guided Practice

1. A factory is making plastic display windows for business calculators. The windows measure 8.25 centimeters by 1.5 centimeters. How many square centimeters is the area of the display windows?

a. Multiply 8.25 by 1.5 just as you would with whole numbers.

$$825$$
$$\times \quad 15$$
$$\overline{12375}$$

b. Count the decimal places after the decimal point.

8.25 (—— places after decimal point)

× 1.5 (—— places after decimal point)

12375 (—— total places after decimal point)

c. Place the decimal point in the answer.

8.25

× 1.5

12375

Exercises

Write each product, placing the decimal point carefully.

2. 5.8 × 7.5

3. 9.8 × 5.25

4. 4.1 × 8.525

_____

5. 31.4 × 96.3

6. 67.4 × 82.1

7. 103.2 × 9.5

_____

8. 3.53 × 5.8

9. 49.3 × 5.2

10. 56.5 × 10.1

_____

11. 0.631 × 1.98

12. 0.047 × 0.024

13. 8.7 × 0.0025

_____

14. Marta received pledges for $21.50 per kilometer, and she walked 8.9 kilometers. How much money did she earn for charity?

_____

Round to the nearest cent where needed. You may use a calculator if you wish.

15. At $4.30 per meter, what is the cost of 10.75 meters of lumber?

_____

16. Jong makes $6.95 for an hour of overtime work. How much does he make for 12.5 hours of overtime work? _____

17. How far can you travel on 14.3 gallons of gasoline if your car averages 22.5 miles per gallon? _____

18. It costs $0.95 per hour to use a 200-watt light bulb. How much does it cost to use the bulb for 3.25 hours? _____

19. Find the area of a tabletop that is 62.5 inches long and 32.25 inches wide. _____

20. Luann wants to tile her floor. The tile she picks measures 1.75 meters on every side.

a. What area will she be able to cover with one tile?

_____

b. What area will 40 tiles cover? _____

21. Aaron worked as a plumber for $25.95 an hour. How much did he earn for 37.5 hours of work last week? _____

22. How much glass is needed to cover the front of a photograph that measures 6.9 inches by 9.2 inches? _____

23. Use a calculator to do Exercises 2 to 13. As an experiment, see which method is faster: entering the decimal points when you multiply, or multiplying without entering decimal points, then inserting the decimal point yourself by quickly figuring out the total number of digits that should come after the decimal point.

_____

_____

_____

24. Predict the number of decimal places in the product of 0.082 and 0.986; then compute the product. Was your prediction correct? Why or why not?

_____

_____

_____

Vocabulary

quotient: the answer in a division problem

ESTIMATING DECIMAL QUOTIENTS

Example 1

Jerry works at the local minor league ball park on weekends. He hears the manager say that the team has taken in $27,600 for today's game. Jerry knows that most tickets were $6.75. He wants to estimate how many people are at the game.

First, he rounds *$27,600* to *$28,000*.

Next, he rounds *$6.75* to *$7*.

Then he mentally divides the whole numbers.

$$\$28{,}000 \div \$7 = \$4{,}000$$

There are about 4,000 people at the game.

Example 2

What is $453 divided by $8.90? Estimate the quotient.

Knowing multiplication facts for 9 will help you choose numbers that are close to the given numbers and are easy to divide.

$453 is close to $450.

$8.90 is close to $9.

$450 ÷ $9 = $50

So, $453 divided by $8.90 is about $50.

Guided Practice

1. At the ballpark, Jerry sold hats for $3.95. He took in $476. Approximately how many hats did he sell?

   a. Round $3.95 to the nearest whole number. *$4*

   b. Round $476 to a number you can easily divide by 4. _____

   c. Divide mentally to find the answer. _____

Estimate each quotient.

2. 715 ÷ 6.8   **3.** $512 ÷ $24.88   **4.** $892 ÷ 4.9

_____   _____   _____

5. 14,876 ÷ 15.4   **6.** 811 ÷ 3.9   **7.** 49,010 ÷ 6.7

_____   _____   _____

8. 98,000 ÷ 10.1   **9.** 15.30 ÷ 4.53   **10.** $65.20 ÷ 10.3

_____   _____   _____

11. A school play took in $423. Tickets were $2.25 each.

Estimate how many students attended the play. _____

12. One evening at the ball park, a local charity asked for donations. There were about 1,800 people at the game and $920 was collected. Estimate the average amount that each person gave. _____

13. The Montoya family drove 1,784 miles and used 63 gallons of gas.

    a. About how many miles did they travel on each gallon of gas?

    _____

    b. On average, they spent $1.20 for each gallon of gas. Estimate the total amount they spent on gas.

    _____

Application

COOPERATIVE

LEARNING

14. Work with a partner to test which is faster, estimation or a calculator. For each of the following decimal division problems, one partner will estimate the quotient while the other partner finds the exact quotient using a calculator. Write your answers on a piece of paper. Who finished first? Compare your estimates to the exact answers to be sure they are close.

    a. 31.2 ÷ 5.1   **b.** 99.9 ÷ 20.2

    c. 49.05 ÷ 6.95   **d.** 13.89 ÷ 1.9

    e. 35.5 ÷ 3.6

DIVIDING BY POWERS OF 10

When you divide a number such as 653 by 10 and by 100, you get the following results.

$$
\begin{array}{r}
65.3 \\
10\overline{)653.0} \\
\underline{60} \\
53 \\
\underline{50} \\
30
\end{array}
$$

$$
\begin{array}{r}
6.53 \\
100\overline{)653.00} \\
\underline{600} \\
530 \\
\underline{500} \\
300
\end{array}
$$

Only the position of the decimal point differs in the two answers.

To divide a number by 10, move the decimal point one place to the left. Notice that 10 has one zero. To divide a number by 100, move the decimal point two places to the left. Notice that 100 has two zeros. You can use this pattern to predict that dividing by 1,000 will result in the quotient 0.0653.

You now have a rule for dividing a number by a power of 10: move the decimal point to the left as many places as there are zeros in the power of 10.

$$2645.9 \div 100 = 2\ 6\ 4\ 5.9 = 26.459$$

Write zeros to the left of the number if you do not have enough places to move the decimal point.

$$97.2 \div 10,000 = 0.00972$$

Example 1

Suppose you count the pennies in a bank and find there are 643. What is the dollar value of the pennies?

Since every 100 pennies is a dollar, find out how many hundreds there are in 643.

$$643 \div 100 = 6.43$$

Reminder

Recall powers of 10.
$100 = 10^2$; $1,000 = 10^3$; and $10,000 = 10^4$.

So, the value of the pennies is $6.43.

Example 2

In laying out a Walk for Hunger route, Luis finds that a city block in his neighborhood measures 148 meters. He needs to know how many kilometers that is.

Because every 1,000 meters is a kilometer, he needs to know how many thousands are in 148.

$$148 \div 1,000 = 0.148$$

So, a block is 0.148 km long.

Guided Practice

Divide. Place the decimal point in the quotient.

1. $847 \div 100 = \underline{\;8.47\;}$

2. $6,380 \div 10 = \underline{\;638.0\;}$

3. $92.43 \div 1,000 =$

_____

4. $7 \div 10 =$

_____

Exercises

Divide.

5. $3.7 \div 100 =$ _____

6. $0.0482 \div 10 =$ _____

7. $64,500 \div 100 =$ _____

8. $873,600 \div 1,000 =$ _____

9. $0.783 \div 10 =$ _____

10. $27.49 \div 1,000 =$ _____

11. The members of the Neighborhood Youth Club contributed a total of 2,364 pennies for the purchase of stepping stones for the community garden. A construction company offered the stones for $1 each. How many stones could they buy? _____

12. The YWCA pool is 22 m long. Amanda swam 12 lengths. How many kilometers did she swim? _____

Application

13. How does the quotient compare with the dividend when you divide a number by 1? a number greater than 1? a number less than 1?

_____

_____

DIVIDING DECIMALS BY WHOLE NUMBERS

Example 1

Jimmy works as a babysitter for the Ricardo family. One evening, he earned $15.50 for 5 hours of work. He wanted to find his hourly rate of pay. He thought, "I would multiply the number of hours worked times my hourly rate of pay to find my total pay." Then he thought, "Division is the reverse of multiplication, so I should divide to find the hourly rate."

Hours worked × Hourly rate = Total pay

Total pay ÷ Hours worked = Hourly rate

To divide a decimal by a whole number, write a decimal point in the quotient directly above the decimal point in the dividend. Then divide as you would whole numbers.

$$
\begin{array}{r}
3.10 \leftarrow \textbf{Quotient} \\
5\overline{)15.50} \leftarrow \textbf{Dividend} \\
\underline{-15} \\
05 \\
\underline{-5} \\
0
\end{array}
$$

His hourly rate of pay is $3.10.

Example 2

Mrs. Ricardo earned $1,655.40 last month. She worked 31 hours each week. What is her hourly rate of pay?

Find the number of hours she worked last month.

$$31 \times 4 = 124$$

Find her hourly rate of pay.

$$\$1,655.40 \div 124 = \$13.35$$

Her hourly rate of pay is $13.35.

Guided Practice

1. Ashley's father earns $5,099.20 each month. He works 40 hours each week. What is his hourly rate of pay?

a. Find the number of hours he works each month.

40 × 4 = _____

b. Divide the total pay by the number of hours.

_____ ÷ _____ = _____

c. His hourly rate of pay is _____.

Exercises

Divide.

2. 27)$29.43

3. 5)12.065

4. 58)34.8

Find the hourly rate of pay for each total and the time worked. Round quotients to the nearest cent.

5. $81.25 for 25 hours

6. $38.20 for 33 hours

7. $262.20 for 38 hours

_____

_____

_____

8. $1,010.80 a month for 35 hours each week

9. $1,653.12 a month for 41 hours each week

_____

_____

Application

10. Flora worked 44 hours last week. She worked 40 hours at regular pay and 4 hours at time and a half. *Time and a half* means that her hourly rate of pay for each overtime hour was 1.5 times her regular hourly rate. She earned a total of $322. What is her regular hourly rate of pay? _____

11. Explain how you found your answer for Application 10.

_____

_____

_____

DIVIDING A DECIMAL BY A DECIMAL

Example 1

Tyrone and Vincente bought bread for the class sandwich sale. They paid a total of $19.20 for the bread. Each loaf costs $0.80. How many loaves of bread did they buy?

Divide $19.20 by $0.80 to find the number of loaves.

$$\text{Divisor} \rightarrow \overset{\leftarrow \textbf{Quotient}}{\$0.80\overline{)\$19.20}} \leftarrow \textbf{Dividend}$$

To divide a decimal by a decimal, follow these steps.

Step 1   Make the divisor a whole number by multiplying by a power of 10.

Step 2   Multiply the dividend by the same power of 10.

Step 3   Write the decimal point in the answer directly above the decimal point in the new dividend.

Step 4   Divide as you would with whole numbers.

$$
\begin{array}{r}
2. \\
\textbf{Divisor} \rightarrow 0.80.\overline{)19.20.} \\
-16 \\
\hline
32 \\
-32 \\
\hline
0
\end{array}
$$

Move the decimal point two places to the right. Divide 1,920 by 80.

So, Tyrone and Vincente bought 24 loaves of bread.

Example 2

Divide 1.89 by 0.006.

$$
\begin{array}{r}
315. \\
0.006.\overline{)1.890.} \\
-18 \\
\hline
9 \\
-6 \\
\hline
30 \\
-30 \\
\hline
0
\end{array}
$$

Move the decimal points three places to the right. Write zeros in the dividend as needed. Divide 1,890 by 6.

Reminder

To multiply a number by a power of 10, move the decimal point to the right as many places as there are zeros in the power of 10.

Example 3

Divide 0.272 by 3.4.

$$
\begin{array}{r}
0.08 \\
3.4\overline{\smash{)}0.2.72} \\
\underline{-272} \\
0
\end{array}
$$

Move the decimal point, one place to the right. Write zeros in the quotient as needed. Divide 2.72 by 34.

Guided Practice

1. Vered's order of canned soup costs $28.08. The price of soup is $0.78 per can. How many cans of soup did she buy?

   a. Write a division problem. $0.78\overline{\smash{)}28.08}$

   b. How many places to the right will you move the decimal points?

   _____

   c. Divide. She bought _____ cans of soup.

2. Roy's order of milk costs $22.20. The milk is $1.48 a gallon. How many gallons of milk did he buy?

   a. Divide using your calculator. $22.20 \div 1.48$

   b. Roy bought _____ gallons of milk.

Divide.

3. $0.03\overline{)15}$

4. $0.004\overline{)16}$

5. $0.7\overline{)5.6}$

6. $0.12\overline{)7.2}$

7. $3.7\overline{)5.18}$

8. $6.14\overline{)0.921}$

Divide. Use your calculator to check your answer.

9. $\$12.25 \div \$0.49$

10. $\$61.50 \div \$1.23$

11. $\$81.60 \div \$3.40$

_____

_____

_____

12. $\$8.10 \div \$0.09$

13. $\$131.25 \div \$8.75$

14. $\$93.60 \div \$2.60$

_____

_____

_____

15. $\$63 \div \$1.75$

16. $\$40.32 \div \$3.36$

17. $\$175 \div \$17.50$

_____

_____

_____

18. Marcia buys a box of mixed candies for $36. She estimates that there are 900 candies. What is the price for each piece of candy?

_____

COOPERATIVE LEARNING **Work with a partner to study unit prices.**

19. At a local supermarket choose five items each and record the cost of each item and the amount that is contained in the package. For example, a package of hot dogs might cost $2.40 and contain 12 hot dogs. Or a container of yogurt might cost $0.49 and contain 7 ounces of yogurt.

a. Using cost and amount, determine the unit price for each item you've selected. Notice that you may be finding cost per ounce, per gallon, or per piece, depending on what the item is and how it is packaged.

b. Compare unit prices for different brands of the same product. Write your findings below.

_____

_____

_____

DECIMALS AND PERCENTS

Vocabulary

percent: a part compared to one hundred parts

Example 1

The students at Harrison School voted on their favorite kinds of books. Sylvia and Bob counted the votes.

Biography   20

Science fiction 30

Novel   18

History   12

"These numbers won't mean much to people," Sylvia said.

"You're right," said Bob. "Let's figure out what **percent** of students voted for each category."

First, find the total number of votes.

$$20 + 30 + 18 + 12 = 80$$

Then, to find what percent of the students voted for biography, divide 20 by the total number of votes.

$$
\begin{array}{r}
.25 \\
80\overline{)20.00} \\
-\underline{160} \\
400 \\
-\underline{400} \\
0
\end{array}
$$

Write zeros in the dividend so you can divide until there is a zero remainder.

To change a *decimal to a percent*, move the decimal point two places to the right and write a percent sign.

$$0.25. = 25\%$$

Biography was selected by 25% of the students.

Example 2

Change 5% to a decimal.

5% means 5 out of 100, or $5 \div 100 = 0.05$

To change a *percent to a decimal*, drop the percent sign and move the decimal point two places to the left. Write zeros as needed.

$$5\% = 0.05$$

1. Change 0.275 to a percent.

   a. Move the decimal point two places to the right.

   0.27.5

   b. Write 0.275 as a percent. _____

Exercises

Write each decimal as a percent.

2. 0.25

3. 0.99

4. 0.01

_____   _____   _____

5. 0.30

6. 0.375

7. 0.5

_____   _____   _____

Write each percent as a decimal.

8. 20%

9. 52%

10. 3%

_____   _____   _____

11. 75%

12. 1%

13. 12.5%

_____   _____   _____

Application

14. Use a bag of small candies such as sour balls. Before you open the package, guess what percent of the pieces will be what color. Then calculate the percent of each color by counting the number of each color and dividing this by the total number of candies.

_____

 # MIXED APPLICATIONS OF DECIMALS

Reminder

When you withdraw money
from a bank account,
you subtract that amount
from your account balance.

Example 1

Suzanne opened her checking account with $175. She
then paid a check to Sights and Sounds for $17.98 for a
CD on February 3. Next, she paid $21.56 to the Galleria
Shop on February 3. She paid $45 to the Woolen Store
on February 5. Finally, she deposited a paycheck for
$67.98 in her account on February 6. What is Suzanne's
account balance?

To find out, Suzanne wrote these transactions in her
checkbook register below, showing the new balance after
each transaction. She subtracted each withdrawal and
added her deposit.

RECORD ALL CHARGES OR CREDITS THAT AFFECT YOUR ACCOUNT							
NUMBER	DATE	DESCRIPTION OF TRANSACTION	PAYMENT	FEE IF ANY	DEPOSIT	BALANCE	
						175	00
101	2/3	Sights & Sounds	-17 98			17	98
						157	02
102	2/3	Galleria Shop	-21 56			21	56
						135	46
103	2/5	The Woolen Store	-45 00			45	00
						90	46
	2/6	Paycheck			+67 98	67	98
						158	44

Her balance is $158.44.

Example 2

Mario has $9 in his pocket when he stops at a fast-food
restaurant. He wants a hamburger and coleslaw for
$1.79, french fries for $0.59, a soda for $0.44, and a
sundae for $0.88. But he also wants to see a movie,
which costs $5.25. Does he have enough money?

First, he estimates. "The hamburger costs about $2, the
soda and fries about $1, and the sundae about $1.
That's $4. It's going to be close." Then he takes out a
pencil and adds the prices on a napkin.

$$\$1.79$$
$$.59$$
$$.44$$
$$\underline{.88}$$
$$\$3.70$$

Subtracting $3.70 from $9 leaves $5.30. So Mario has just enough money.

Example 3

A bike that usually sells for $120 is on sale for 15% off. How much will Rolanda save by buying the bike on sale?

To find the amount saved, Rolanda finds 15% of $120.

First, she writes 15% as a decimal.

$$15\% = 0.15$$

Then she multiplies.

$$\$120 \times 0.15 = \$18$$

She will save $18.

Example 4

Oscar's dad bought a used car with 42,780.3 miles on the odometer. He filled the tank with gas. The next time he filled the tank, it took 14.3 gallons of gas, and the odometer read 43,061.7 miles. He asked Oscar to figure how many miles the car gets to a gallon of gas.

First, Oscar figured out the difference between the two odometer readings.

$$43,061.7$$
$$- \ 42,780.3$$
$$281.4$$

Then he divided 281.4 miles by 14.3 gallons to find the number of miles driven for each gallon of gas.

$$143\overline{)281.4.} \rightarrow \overset{19.67}{143\overline{)2814.00}}$$ Write zeros to divide to two decimal places.

Oscar tells his dad the car gets about 19.7 miles to the gallon.

Reminder

To write a percent as a decimal, drop the percent sign and move the decimal point two places to the left.

Guided Practice

1. Francisco has a job using his own car to deliver pizzas. In one week, he drove 998.7 miles and spent $103.65 on gas. What is his gasoline cost per mile?

   a. Divide $103.65 by 998.7

b. Divide to four decimal places. _____

c. Multiply the answer by 100 to find the cents value. _____

d. His gasoline costs _____ per mile, rounded to the nearest tenth.

2. Ellie's transactions are partly entered in her checkbook register below. Complete her entries and calculate her balance after each transaction. Opening balance, $278; check to Sights and Sounds, $45.98, on March 15; deposit, $75, on March 16; payment to Allsport, $76.85, on March 17. Begin numbering checks with Number 201.

	NUMBER	DATE	DESCRIPTION OF TRANSACTION	PAYMENT		FEE IF ANY	DEPOSIT		BALANCE	
			RECORD ALL CHARGES OR CREDITS THAT AFFECT YOUR ACCOUNT						278	00
a.	201	3/15	Sights & Sounds	-45	98				45	98
									232	02
b.		3/16	Deposit				+75	00		
c.	202									
d.										

e. What is Ellie's account balance on March 17? _____

Exercises

Solve each problem.

3. Enter each transaction in the checkbook register and calculate the balance after each transaction. Include the dates of the transactions. Opening balance, $456; check to Corner Clothier for $145.89 on May 1; check to Telephone Co. for $35.67 on May 3; deposit $76.80 on May 4; check to the Food Emporium for $119.45 on May 7. Begin with check Number 301.

a.

NUMBER	DATE	DESCRIPTION OF TRANSACTION	PAYMENT		FEE IF ANY	DEPOSIT	BALANCE	
		RECORD ALL CHARGES OR CREDITS THAT AFFECT YOUR ACCOUNT						
301								

b. What is the account balance on May 7? _____

4. The odometer on the company truck read 53,641.9 miles at 8 A.M. when Shanti began a trip to pick up merchandise at a distributor across the state. When she arrived there at 12:30 P.M., the odometer read 53,898.4 miles. How many miles per hour did she average? _____

5. An $18.95 video is on sale at a 5% discount. Find the amount saved. _____

6. Ralph spent $714 on flooring for the kitchen. If the floor covering costs $8.95 a square foot, estimate the area of the kitchen. _____

Application

 Solve.

7. On a given day, one U.S. dollar was worth 1.28 Canadian dollars, 780.65 South Korean won, 126.65 Japanese yen, 25.92 Indian rupees, and 1,232 Italian lira. Use these exchange rates and a calculator to find the price of the items in the table and three additional items of your choice in the countries listed.

		United States	Canada	South Korea	Japan	India	Italy
a.	Sport watch	$34.00					
b.	In-line skates	$117.00					
c.	Wallet	$29.00					
d.							
e.							
f.							

Find the value of the 5 in each number.

1. 79,061.05

2. 36.005

3. 1,100.157

_____   _____   _____

Write in words.

4. 20.003 _____

5. 73.12 _____

Use the number line below to locate and label the following points.

6. 3.6   **7.** 3.3   **8.** 2.5

```
←+++++++++++++++++++++++++++++++++++++++++++++→
   0        1        2        3        4
```

Write the numbers in order from least to greatest.

9. 67.015; 67,015; 67.025   _____

10. 345.01; 345.009; 345.901 _____

Compare. Write <, >, or =.

11. 89.05 _____ 89.049   **12.** 1.1 _____ 1.099

13. As a carpenter's assistant, Carmelita was asked to write these measurements. How would she write each amount in standard form?

a. twenty-seven thousandths _____

b. three and thirteen hundredths _____

c. nine and eight tenths _____

d. three hundred twenty and fifty-four hundredths _____

5-8 CUMULATIVE REVIEW

Round each number to the underlined place value.

1. 6.<u>2</u>7 _____

2. 75.01<u>2</u>3 _____

3. 512.3<u>9</u>5 _____

4. $29<u>9</u>.87 _____

5. Round $946.52 to:

    a. the nearest dollar. _____

    b. the nearest ten cents. _____

    c. the nearest hundred dollars. _____

Write each decimal as a fraction.

6. 0.20 _____ 7. 0.003 _____ 8. 0.5 _____ 9. 0.75 _____

Write each fraction as a decimal.

10. $\frac{4}{5}$ _____ 11. $\frac{9}{10}$ _____ 12. $\frac{1}{3}$ _____ 13. $\frac{3}{20}$ _____

Estimate each sum or difference.

14. $12.7 - 3.1$ 15. $90.25 + 3.90$ 16. $100.5 - 1.2$ 17. $16.11 + .89$

_____ _____ _____ _____

18. Isabel goes shopping with $20. She plans to purchase a T-shirt for $8.75, a book for $7.95, and two magazines for $1.50 each. Estimate whether or not she has enough money.

_____

Estimate each sum. Then line up the decimal points and add.

1. Add: 29.56, 8.123, 51.9

Estimate: _____

Answer: _____

2. Add: $18.95, $108, $75.63

Estimate: _____

Answer: _____

3. Add six and two hundred five thousandths plus thirty-seven and eight tenths. _____

Find each sum. Write your answer in meters.

4. Add: 47.8 m + 2.6 km + 103.5 m _____

5. Add: 321 cm + 18.4 m + 175.2 cm _____

Estimate the difference. Then line up the decimal points and subtract.

6. Subtract 48.3 from 295.

Estimate: _____

Answer: _____

7. From $50 subtract $8.95.

Estimate: _____

Answer: _____

8. Maria went to the store. She purchased socks for $3.19, a T-shirt for $7.63, and jeans for $19.04.

a. How much did she spend? _____

b. How much change would she receive from $30? _____

Find each difference. Write your answer in meters.

9. From 2 m subtract 45 cm. _____

10. Subtract 15.6 m from 0.8 km. _____

11. Jay works as a carpenter's apprentice. He cuts 2-by-4s into standard 1-meter lengths. He has a pile of wood with the following measurements.

120 cm   200 cm   90 cm   350 cm   400 cm

How many 1-meter boards can he cut from these pieces? _____

1. The cost to pave the playground at a nearby park is $4.15 per square yard. The playground is 30.4 yards long and 14.8 yards wide. Round all decimals to the nearest whole number.

   a. Estimate the area to be paved. _____

   b. Estimate the cost of this paving project. _____

2. Estimate the area of a rectangle of length 31.7 ft and width 14.9 ft.

   _____

Multiply.

3. $24.6 \times 100 =$ _____

4. $3.73 \times 10 =$ _____

5. $0.00085 \times 1,000 =$ _____

In the spreadsheet below, what numbers will appear in cells D1, D2, D3, and D4?

		A	B	C	D	
6.	1	5.85	8	5	A1*B1*C1	_____
7.	2	4.50	7.5	3	A2*B2*C2	_____
8.	3	9.20	5	4	A3*B3*C3	_____
9.	4	6.70	8	5	A4*B4*C4	_____

Multiply.

10.	4.279	11.	10.02	12.	13.52
	× 3.15		× 0.5		× 23.22

13. A string is 14.2 centimeters long. How long are 15 strings?

   _____

14. A gallon of gasoline costs $1.24. How much will 12.6 gallons cost? Round your answer to the nearest cent.

   _____

Estimate each quotient.

1. $38 \div 6.2 =$ _____   **2.** $\$3.09 \div 24.8 =$ _____   **3.** $19.72 \div 4.9 =$ _____

Divide.

4. $5.64 \div 100 =$ _____   **5.** $12.3 \div 10 =$ _____   **6.** $278.9 \div 1{,}000 =$ _____

7. $9\overline{)11.52}$ **8.** $35\overline{)164.5}$ **9.** $78\overline{)8.814}$

10. $0.4\overline{)26.4}$ **11.** $0.53\overline{)6.519}$ **12.** $0.108\overline{)18.36}$

Solve.

13. Rachel bought a package of four batteries for $2.69. Estimate how much she paid for each battery. _____

14. If $20.25 is divided among nine friends, how much money does each person get? _____

15. Pierre paid $1.09 for a 3.5-ounce bar of soap. What was the price per ounce? _____

Write each decimal as a percent.

1. 0.85   **2.** 0.37   **3.** 0.09

_____   _____   _____

Write each percent as a decimal.

4. 22%   **5.** 41%   **6.** 6%

_____   _____   _____

Solve each problem.

7. Isaac purchased 3 quarts of ice cream at $3.24 each.
How much did he pay for the ice cream? How much change did he
receive from $20? _____

8. Lilly worked for 34 hours and earned $282.54. What was her hourly
rate of pay? _____

9. Omar worked for 39 hours at $6.50 an hour. How much did he earn?

_____

10. A store got a book shipment costing $187.20. The books were $3.90
each. How many books did the store buy? _____

11. Adam bought items that cost $45.67, $18.29, and $7.06. What is his
change from $100? _____

12. Find the savings on a $139 CD player if the discount is 20%.

_____

ANSWER KEY

LESSON 1 (pp. 2 – 3)

  1. a. 9   **b.** 9 × 1,000,000 or 9,000,000   **c.** 3

  d. 3 × 10 or 30   **e.** 5   **f.** 5 × 0.001 or 0.005

  3. 4 × 0.1 or 0.4   **5.** 9 × 0.01 or 0.09

  7. 9 × 10 or 90   **9.** 5 × 0.1 or 0.5

  11. The greatest number you can write is 9,876,543.210. Sample: I wrote the greatest digit in the place with the greatest value.

LESSON 2 (pp. 4 – 5)

  1. a. ten   **b.** five hundredths

  c. ten and five hundredths

  3. one and five hundredths

  5. 324,000   **7.** 9.007

LESSON 3 (pp. 6 – 7)

  1.

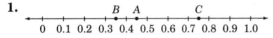

  3.

  5.   1.3

  7.   0.75

  9.

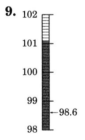

  11. 3,005.20   **13.** 25.155

LESSON 4 (pp. 8 – 9)

  1. a. yes   **b.** yes   **c.** yes   **d.** >   **e.** >

  f. 450.09, 450.1, 450.17

  3. 32.98, 39.09, 39.6   **5.** 5.75, 15.75, 57.5

  7. 429.01, 429.34, 429.6

  9. <   **11.** =

LESSON 5 (pp. 10 – 11)

  1. a. 4   **b.** 9   **c.** greater than 5   **d.** 155.75

  3. $67.10   **5.** 99.0192   **7.** $753.00

  9. a. – b. Check students' answers.

  c. Rounding to the nearest dime is more accurate.

LESSON 6 (pp. 12 – 13)

  1. a. $\frac{25}{100}$ **b.** $\frac{1}{4}$

  3. $\frac{1}{10}$ **5.** $\frac{1}{100}$ **7.** $\frac{3}{8}$ **9.** $\frac{1}{20}$ **11.** $\frac{123}{1,000}$

  13. $\frac{500}{1,000} = \frac{1}{2}$ **15.** Check students' charts.

LESSON 7 (pp. 14 – 15)

  1. denominator   **3.** 0.83 $\frac{1}{3}$

  5. 0.625   **7.** 0.16 $\frac{2}{3}$ **9.** 0.66 $\frac{2}{3}$ **11.** $\frac{4}{5}$

  13. Check students' work.

LESSON 8 (pp. 16 – 17)

  1. a. $20   **b.** $15   **c.** $22   **d.** $57   **e.** no

  3. 57   **5.** 101   **7.** $36   **9.** $26; no

LESSON 9 (pp. 18 – 19)

  1. a. 3.800, 9.356, 4.070   **b.** 17.226 cm

  3. $1805; $1805.03   **5.** 1035; 1034.584

  7. $3.27   **9.** $8.16

LESSON 10 (pp. 20 – 21)

  1. a. 1,000; 4,250   **b.** 4,250; 4,385 m

  c. 4,385 m

  3. 146 cm   **5.** 213.5 cm

  7. 7.6 km   **9.** 51.5 m

LESSON 11 (pp. 22 – 23)

  1. a. $405.08   **b.** $99.85   **c.** $305.23

  3. 37.8   **5.** $349; $348.15

  7. 69; 68.71   **9.** Check students' work.

LESSON 12 (pp. 24 – 25)

  1. a. 3,400   **b.** 3,400 − 800 = 2,600

  c. 2,600 m, 2.6 km

  3. 195.4 m   **5.** 1,270 m

  7. Check students' work.

LESSON 13 (pp. 26 – 27)
   1. a. $3, 25, 15   **b.** 375 sq yd   **c.** $1,125
   3. $8 \times 8 = 64$ sq m
   5. $400 \times 500 = 200,000$ sq mi
   7. $7 \times 7 = 49$ sq m

LESSON 14 (pp. 28 – 29)
   1. 37.1   **3.** 12,800   **5.** 30
   7. 678   **9.** 3.6   **11.** $290
   13. Answers may vary. Students should
      mention that these numbers are easier to
      read and write.

LESSON 15 (pp. 30 – 31)
   1. a. 0.0028   **b.** 0.186   **c.** 37.20
   3. 0.1713   **5.** 29.1   **7.** 112.256   **9.** 46.8   **11.** 21

LESSON 16 (pp. 32 – 35)
   1. a. 12375   **b.** $2 + 1 = 3$ **c.** 12.375
   3. 51.45   **5.** 3,023.82   **7.** 980.4   **9.** 256.36
   11. 1.24938   **13.** 0.02175   **15.** $46.23
   17. 321.75 miles   **19.** 2,015.625 sq in.
   21. $973.13   **23.** Answers will vary.

LESSON 17 (pp. 36 – 37)
   1. a. $4   **b.** $480   **c.** 120 hats
   For Exercises 2–14, estimates may vary.
   3. $20   **5.** 1,000   **7.** 7,000   **9.** 3
   11. 200 students
   13. a. 30 miles   **b.** $72

LESSON 18 (pp. 38 – 39)
   1. 8.47   **3.** 0.09243   **5.** 0.037
   7. 645   **9.** 0.0783   **11.** 23
   13. It is the same.   **15.** It is greater.

LESSON 19 (pp. 40 – 41)
   1. a. 40×4; 160
      b. $5,099.20 ÷ 160; $31.87   **c.** $31.87
   3. 2.413   **5.** $3.25   **7.** $6.90   **9.** $10.08
   11. Possible answer: 4 hours at 1.5 times her
      regular hourly rate is the same as 6
      (1.5×4) hours at regular rate;
      $40 + 6 = 46$ hours worked;
      $322 ÷ 46 = $7

LESSON 20 (pp. 42 – 45)
   1. a. $0.78\overline{)28.08}$ **b.** 2   **c.** 36   **3.** 500   **5.** 8
   7. 1.4   **9.** 25   **11.** 24   **13.** 15   **15.** 36   **17.** 10
   19. Check students' work.

LESSON 21 (pp. 46 – 47)
   1. a. 27.5   **b.** 27.5%
   3. 99%   **5.** 30%   **7.** 50%
   9. 0.52   **11.** 0.75   **13.** 0.125

LESSON 22 (pp. 48 – 51)
   1. a. $103.65; 998.7   **b.** 0.1037
      c. 10.37   **d.** 10.4¢ per mi
   3. a.

RECORD ALL CHARGES OR CREDITS THAT AFFECT YOUR ACCOUNT						
NUMBER	DATE	DESCRIPTION OF TRANSACTION	PAYMENT	FEE IF ANY	DEPOSIT	BALANCE 456 00
301	5/1	Corner Clothier	-145 89			145 89 / 310 11
302	5/3	Telephone Co.	-35 67			35 67 / 274 44
	5/4	Deposit			+76 80	76 80 / 351 24
303	5/7	Food Emporium	-119 45			119 45 / 231 79

   b. $231.79
   5. about 95¢
   7.

		United States	Canada	South Korea	Japan	India	Italy
a.	Sport Watch	$34.00	43.52	26,542.1	4,306.1	881.28	41,888
b.	In-line Skates	$117.00	149.76	91,336.05	14,818.05	3,032.64	144,144
c.	Wallet	$29.00	37.12	22,638.85	3,672.85	751.68	35,728

   d. -f. Check students' work.

CUMULATIVE REVIEW (L1–L4) (p. 52)

1. 5×0.01 or 0.05

3. 5×0.1 or 0.5

5. seventy three and twelve hundredths

7.

9. 67.015, 67.025, 67,015   **11.** 89.05 > 89.049

13. a. 0.027   **b.** 3.13   **c.** 9.8   **d.** 320.54

CUMULATIVE REVIEW (L5–L8) (p. 53)

1. 6.3   **3.** 512.40 or 512.4

5. a. $947   **b.** $946.50   **c.** $900

7. $\frac{3}{1,000}$ **9.** $\frac{3}{4}$ **11.** 0.9   **13.** 0.15

15. about 94   **17.** about 17

CUMULATIVE REVIEW (L9–L12) (p. 54)

1. about 90; 89.583   **3.** 44.005

5. 23.362 m   **7.** about $40; $41.05

9. 1.55 m   **11.** 11 boards

CUMULATIVE REVIEW (L13–L16) (p. 55)

1. a. 450 sq yd   **b.** $1,800

3. 2,460   **5.** 0.85   **7.** 101.25

9. 268.00   **11.** 5.01   **13.** 213 cm, or 2.13 m

CUMULATIVE REVIEW (L17–L20) (p. 56)

1. 6   **3.** 4   **5.** 1.23   **7.** 1.28   **9.** 0.113

11. 12.3   **13.** about 70¢   **15.** 31¢

CUMULATIVE REVIEW (L21–L22) (p. 57)

1. 85%   **3.** 9%   **5.** 0.41   **7.** $9.72; $10.28

9. $253.50   **11.** $28.98